Guidance

CW01276512

How to use this book:

This workbook contains over 100 questions specifically designed to advance your child's place value knowledge.
Our years of experience have enabled us to put together the perfect balance of fluency, reasoning questions and vocabulary checks.

Each page is dedicated to one of the current curriculum objectives.

You will find one or two pages of fluency questions. These questions help develop number sense and you will see various questions framed in different ways in order for your child to secure their understanding of the objective.
This means your child will not just be memorising facts, but have a real conceptual understanding.

There will then be a page of reasoning questions.
Reasoning will allow your child to explore these objectives at a deeper level and show if your child has truly mastered these concepts.
It requires children to use mathematical vocabulary, explore trial and error and explain how they have reached their answers.

There will be a space for your child to write their answers down.
If your child prefers to explain this verbally, this is equally acceptable.

After each page is completed, your child can colour in the stars at the top of the page to show how many were correct and they can also use the section on page 28 to colour in. This is a great assessment tool to check the progress and identify any gaps in their knowledge.

For more worksheets and advice- please visit our ever growing and popular website:

www.masterthecurriculum.co.uk

These workbooks are dedicated to Tia, Leanna and Malachi.

MTC Publications

Contents

My Place Value Vocabulary

Place Value

The value of a digit, depending on its position.

For example- the numbers 213, 32, and 25 all have the number 2 in it but the place value of 2 is different in all of them.

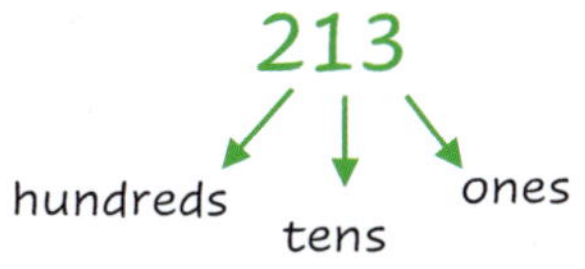

Digit

Any of the ten numbers:
0, 1, 2, 3 ,4, 5, 6, 7 ,8 ,9

The number 52 has two digits.

Tens and Ones

A two-digit number has tens and ones.

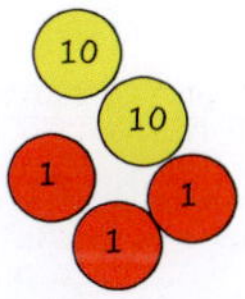

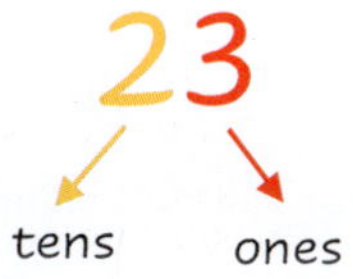

Tens	Ones
2	3

Ten Frames

These can help us count and see the tens easily.

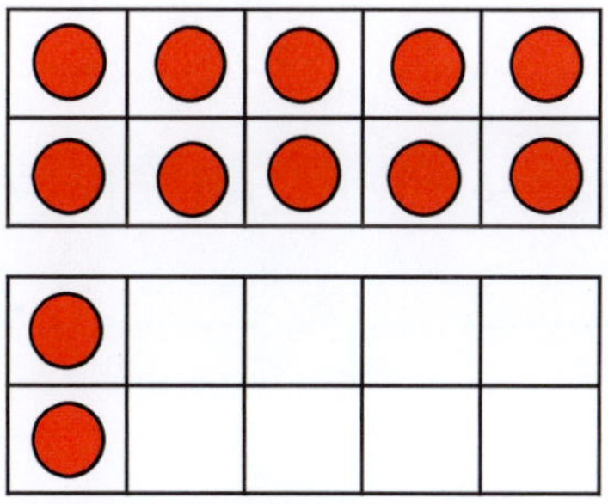

Numeral

A numeral is a **symbol or name** that stands for a number.

For example: 7, ten, 15 and eleven are all numerals.

Number Track

A line of numerals, normally in a pattern.

10	9	8	7	6	5	4	3	2	1	0

0	1	2	3	4	5	6	7	8	9	10

ten	nine	eight	seven	six	five	four	three	two	one	zero

My Place Value Vocabulary

Less / Fewer / Fewest

A smaller quantity or amount.

More / Greater / Greatest

A larger quantity or amount.

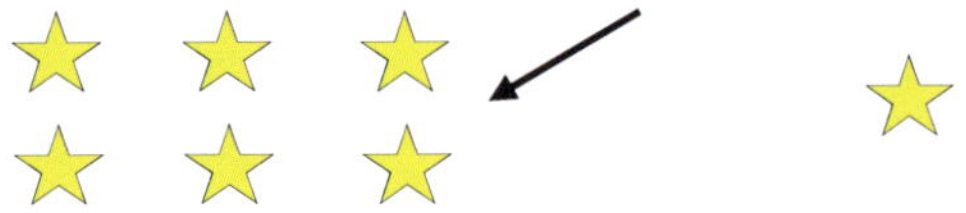

Compare

Looking at the difference between numbers.

Is one greater than the other?

Are they equal to each other?

Representation

Pictorial representation - we can use pictures in maths to stand for a number.

These pictures all represent the number 21.

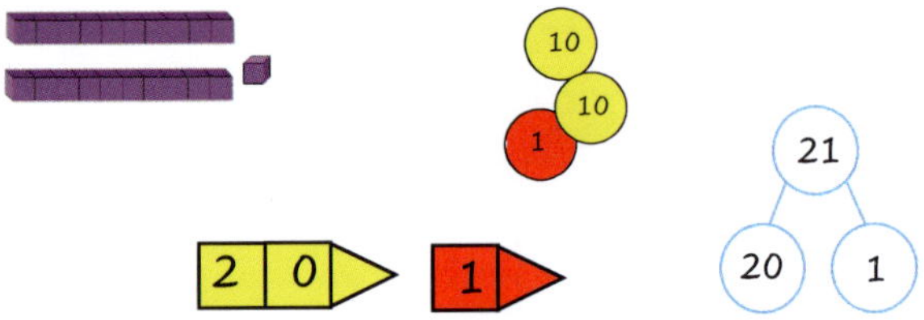

Inequality Symbols

We can use these symbols to tell us if a number is greater than or less than another number.

less than	equal	greater than
<	=	>
4 < 20	**2 tens = 20**	**2 > 0**

Partition

To split / separate / divide numbers into smaller parts. This can make calculations easier.

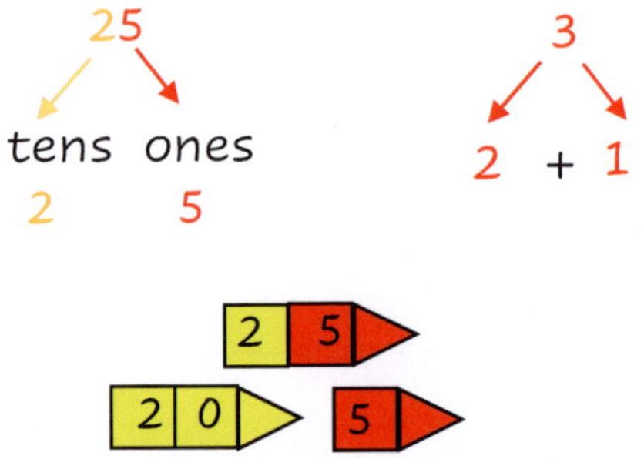

My Place Value Vocabulary

Part-Whole Model

A model that shows parts of a number and the whole number.

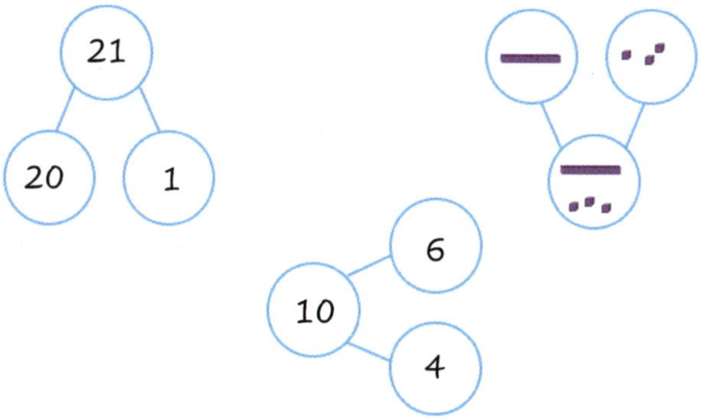

Equal

=

The same amount.

 = 4

The children have an **equal** amount of apples.

Before

6, 7, 8, 9

The number <u>before</u> 7 is 6.

What number comes **before** 24?

20, 21, 22, 23, 24, 25

After

6, 7, 8, 9

The number <u>after</u> 7 is 8.

What number comes **after** 89?

86, 87, 88, 89, 90, 91

Next

Similar to the word after.

0, 1, 2, 3...

What comes **next?**

The number 4 is **next**.
0, 1, 2, 3, **4**

Between

The numbers between 20 and 25 are 21, 22, 23 and 24.

20, 21, 22, 23, 24, 25

The number between 42 and 44 is 43.

41, 42, 43, 44, 45, 46

My Place Value Vocabulary

Sequence

A sequence is a list of numbers, shapes or objects in an order. Sequences do not have to have a pattern, but they normally do.

2, 4, 6, 8, ...

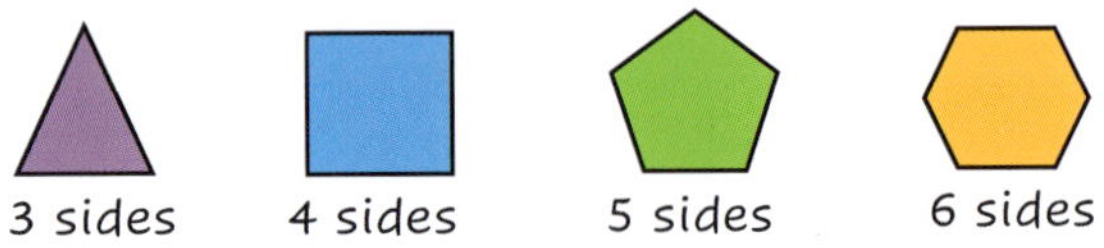

First, I brush my teeth. Next, I have breakfast. Then I get dressed.

Pattern

Like a sequence. It is a list of numbers, shapes or objects in an order, but it follows a rule and is repeated.

2, 4, 6, 8, ...

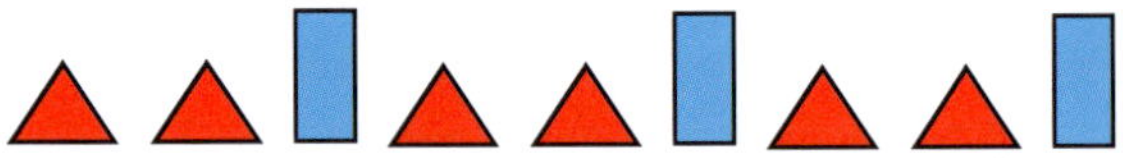

Even Numbers

Even numbers can be shared equally when they are whole. You can spot even numbers visually.

Even numbers end in 0,2,4,6 and 8.

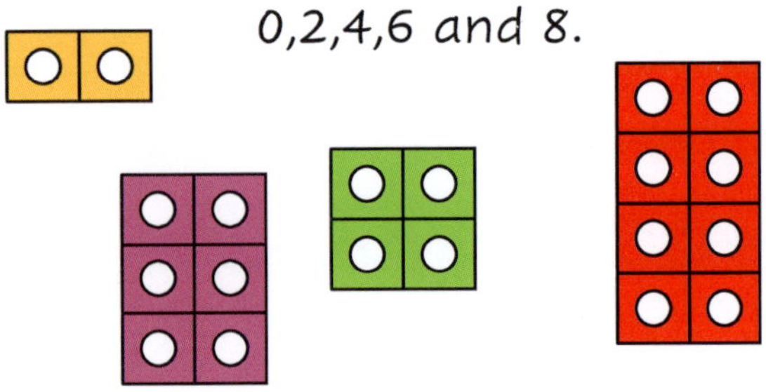

Odd Numbers

Odd numbers can not be shared equally when they are whole. You can spot odd numbers visually.

Odd numbers end in 1,3,5,7 and 9.

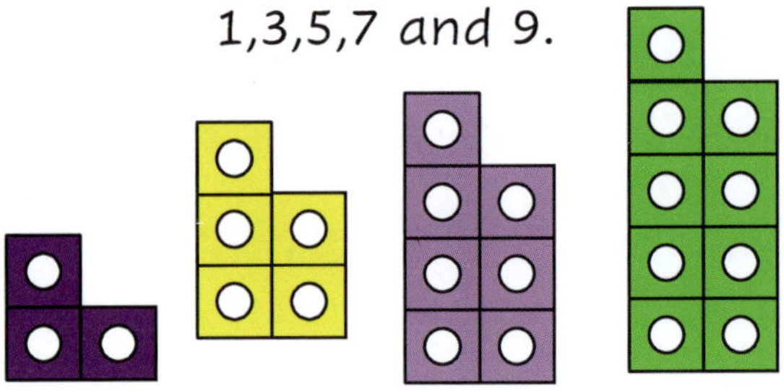

Hundreds

You will learn up to the number 100 and sometimes will count over 100.

Numbers over 99 have three digits.

Hundreds	Tens	ones
1	0	0

Columns and Rows

Columns go down. ⇩
Rows go across. ⇨

These number cards are in a **row**.

0 1 2

These number cards are in a **column**.

0
1
2

Lesson 1
Count in steps of 2 from 0

Use this 100 square grid to practise counting in twos.

What do you notice about all of your numbers?

They are all even.
This means that they always end in 0, 2, 4, 6 or 8.

When you feel confident, practise your skill on pages 12-13.

1	2	3	4	5	6	7	8	9	10
11	12	13	14	15	16	17	18	19	20
21	22	23	24	25	26	27	28	29	30
31	32	33	34	35	36	37	38	39	40
41	42	43	44	45	46	47	48	49	50
51	52	53	54	55	56	57	58	59	60
61	62	63	64	65	66	67	68	69	70
71	72	73	74	75	76	77	78	79	80
81	82	83	84	85	86	87	88	89	90
91	92	93	94	95	96	97	98	99	100

Practise counting in twos until you do not need this 100 square grid.
Can you start with any multiple of two?

Lesson 2
Count in steps of 3 from 0

Use this 100 square grid to practise counting in threes.

Visualising a pattern can help you remember counting in threes.

When you feel confident, practise your skill on pages 14-15.

1	2	3	4	5	6	7	8	9	10
11	12	13	14	15	16	17	18	19	20
21	22	23	24	25	26	27	28	29	30
31	32	33	34	35	36	37	38	39	40
41	42	43	44	45	46	47	48	49	50
51	52	53	54	55	56	57	58	59	60
61	62	63	64	65	66	67	68	69	70
71	72	73	74	75	76	77	78	79	80
81	82	83	84	85	86	87	88	89	90
91	92	93	94	95	96	97	98	99	100

Practise counting in threes until you do not need this 100 square grid.
Can you start with any multiple of three?

Lesson 3

Count in steps of 5 from 0

Use this 100 square grid to practise counting in fives.

What do you notice about all of your numbers?

They always end in 5 or 0.

1	2	3	4	5	6	7	8	9	10
11	12	13	14	15	16	17	18	19	20
21	22	23	24	25	26	27	28	29	30
31	32	33	34	35	36	37	38	39	40
41	42	43	44	45	46	47	48	49	50
51	52	53	54	55	56	57	58	59	60
61	62	63	64	65	66	67	68	69	70
71	72	73	74	75	76	77	78	79	80
81	82	83	84	85	86	87	88	89	90
91	92	93	94	95	96	97	98	99	100

When you feel confident, practise your skill on pages 16-17.

Practise counting in fives until you do not need this 100 square grid. Can you start with any multiple of five?

Lesson 4

Count in tens from any number, forward and backward

You will learn how to count in tens forwards and backwards from any number.

If we add 1 ten to a number, the tens digit increases by 1.

If we go back 1 ten, then tens digit decreases by 1.
You can use a 100 square to help you.
Pick a number and count ten more.
What do you notice?

It is directly underneath the first number and the tens digit has changed

1	2	3	4	5	6	7	8	9	10
11	12	13	14	15	16	17	18	19	20
21	22	23	24	25	26	27	28	29	30
31	32	33	34	35	36	37	38	39	40
41	42	43	44	45	46	47	48	49	50
51	52	53	54	55	56	57	58	59	60
61	62	63	64	65	66	67	68	69	70
71	72	73	74	75	76	77	78	79	80
81	82	83	84	85	86	87	88	89	90
91	92	93	94	95	96	97	98	99	100

Pick a number and count ten less. What did you notice?
The number is directly above it!

When you feel confident, practise your skill on pages 18-20.

This number is 43.

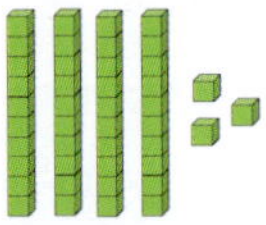

When we add 1 ten, it becomes 53.

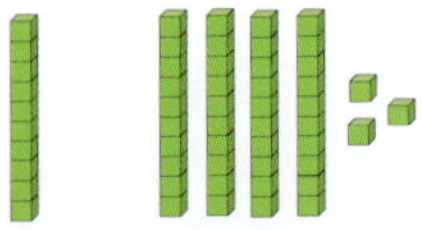

Can you pick any number from 0 – 100 and keep adding ten to it?
Can you count back in tens from a number? Practise.

Lesson 5
Recognise the place value of each digit in a two-digit number

When you see a two-digit number, you will need to understand what each digit represents.

These images help you understand this.

Tens	Ones
2	1

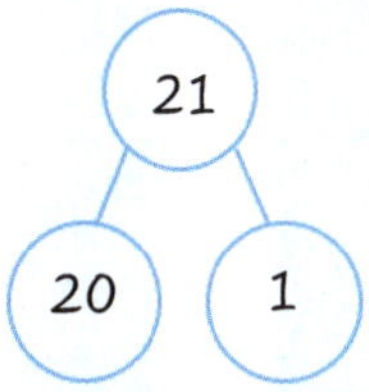

If I had 4 ones and 6 tens, what number would this be?
Some might say 46 but it is 64.
When you read the question properly, you can see it has 4 ones: 
And 6 tens. 

When you feel confident, practise your skill on pages 21-23.

This is the number 64.

Lesson 6
Identify and represent numbers

As you learnt in the vocabulary section, numbers can be represented in different ways.

Look at all of the ways these numbers have been represented.

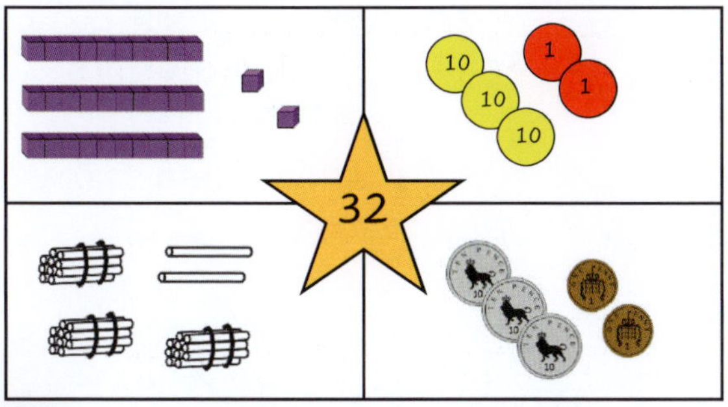

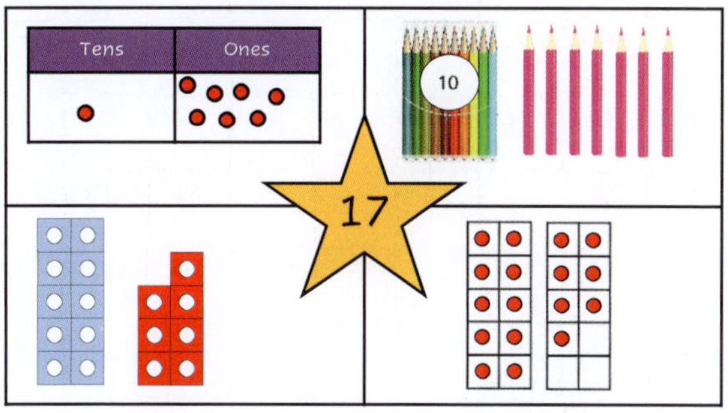

When you feel confident, practise your skill on pages 24-27.

Can you use equipment or drawings to represent the number 50, 9 and the number 100?

Lesson 7
Compare numbers from 0 up to 100 using <, > and = signs

You'll need to remember what these inequality symbols mean.

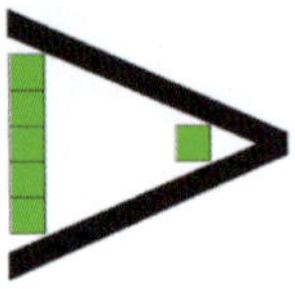

5 is more than 1

> This sign means **more than**.
Use this image to help you remember this.

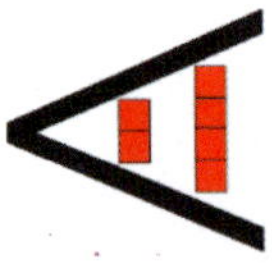

2 is less than 4

< this sign means **less than**.
Use this image to help you remember this.

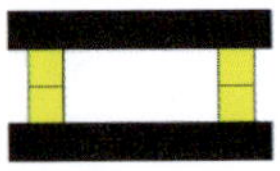

2 is equal to 2

= this sign means equal to.
Use this image to help you remember this.

Read the statements below so you can understand how they work.

8 > 1

8 is more than 1

78 < 87

78 is less than 87

20 > 13

20 is more than 13

99 < 100

99 is less than 100

1 < 100

1 is less than 100

When you feel confident, practise your skill on pages 28-30.

Can you pick two numbers and add the inequality symbols? Swap the numbers around. What do you have to do with the inequality symbols? They will also have to change!

Lesson 8
Order number from 0 up to 100

You will have to order numbers up to 100.

You might be asked to order numbers from smallest to biggest, or biggest to smallest.

56	71	78	87

Smallest to Biggest

39	25	18	6

Biggest to Smallest

When looking at your numbers, ask yourself, does it have one digit or two digits?

One digit numbers are always smaller than 2 digits numbers.

Sometimes, you are given numbers that look similar to each other like these:

47 **74** **17** **41** **14**

Step 1: Check your tens digits. The one with the lowest tens digit will be part of the smallest numbers.

What happens if you have numbers with the same tens digit?

17 **14**

Step 2: Check your ones digits. The one with the lowest ones digit will be the smallest.

17 **14** ← Smallest!

When you feel confident, practise your skill on pages 31-32.

Roll a dice and make some 2-digit numbers.
Can you order them from smallest to biggest and biggest to smallest?

Lesson 9

Read and write numbers to at least 100 in numerals and words

Do you recognise all of the numbers in the grid?

Do you know how to read them when they are written as words?

sixteen = 16

twenty-four = 24

forty-eight = 48

ninety = 90

1	2	3	4	5	6	7	8	9	10
11	12	13	14	15	16	17	18	19	20
21	22	23	24	25	26	27	28	29	30
31	32	33	34	35	36	37	38	39	40
41	42	43	44	45	46	47	48	49	50
51	52	53	54	55	56	57	58	59	60
61	62	63	64	65	66	67	68	69	70
71	72	73	74	75	76	77	78	79	80
81	82	83	84	85	86	87	88	89	90
91	92	93	94	95	96	97	98	99	100

1	2	3	4	5	6	7	8	9	10
one	two	three	four	five	six	seven	eight	nine	ten

11	12	13	14	15	16	17	18	19	20
eleven	twelve	thirteen	fourteen	fifteen	sixteen	seventeen	eighteen	nineteen	twenty

20	30	40	50	60	70	80	90	100
twenty	thirty	forty	fifty	sixty	seventy	eighty	ninety	One hundred

When you feel confident, practise your skill on pages 33-35.

Use the list above to help you write the numbers you find tricky.

Some common mistakes that are made:

The number 40 is written as forty, not fourty.
Eight can be tricky at first.
Ninety still has the e on the end of the word.

Lesson 10

Use place value and number facts to solve problems

Using place value facts will help you solve problems.

Some facts that can help you.

If you know numbers that will make 10 (your number bonds) you will know numbers that will make 100.

4 + 6 = 10 **40 + 60 = 100**

2 + 3 = 5 **20 + 50 = 100**

When you know your number bonds for other numbers as well, you can then work with larger numbers in an easier way.

1 + 7 = 8 **10 + 70 = 80**

5 + 3 = 8 **50 + 30 = 80**

Some numbers might be easier to work with than others.
If you have the number 10, how many more do you need to get to 18? It is 8.

If you have the number 10, how many more do you need to get to 12? It is 2.

Understanding that 1 ten is the same as 10 ones will also help you solve problems.

When you feel confident, practise your skill on pages 36-37.

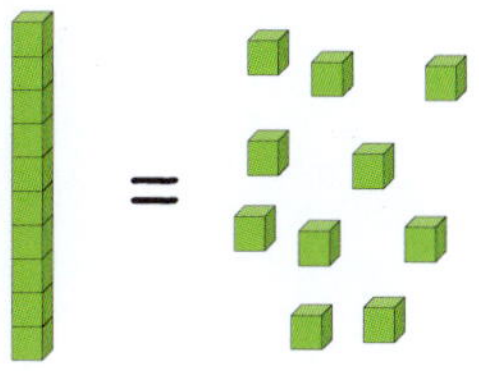

Facts like these can help you solve problems.

Count in steps of 2

1. Count in steps of 2 from the given number.

4, 6, 8, 10, 12, 14 (12 written above)

20, 22, 24, 26, 28, 30

48, 50, 52, 54, 56, 58

2. Complete the sequence.

20, 18, 16, 14, 12

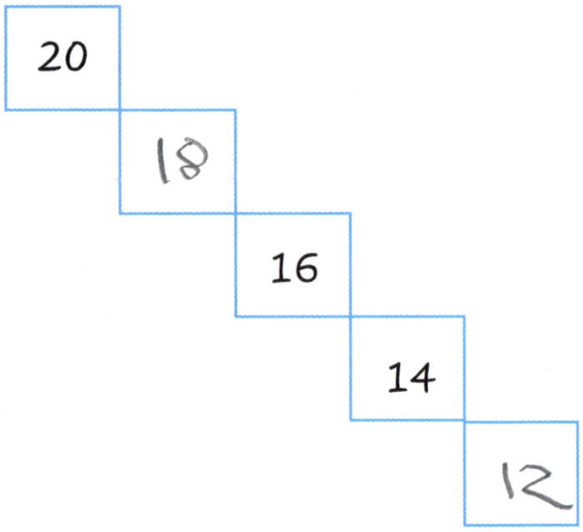

3. Continue the pattern on the caterpillar.

28, 30, 32, 34, 36, 38, 40

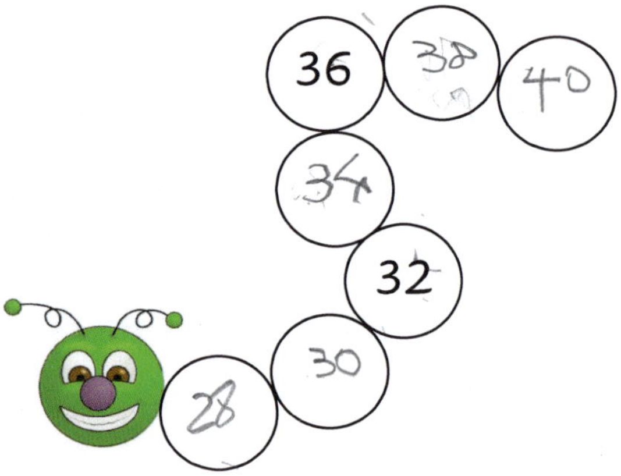

4. What comes before and after in the number sequence? Write the numbers in the box and draw the rest of the sequence.

2, 4, 6, 8

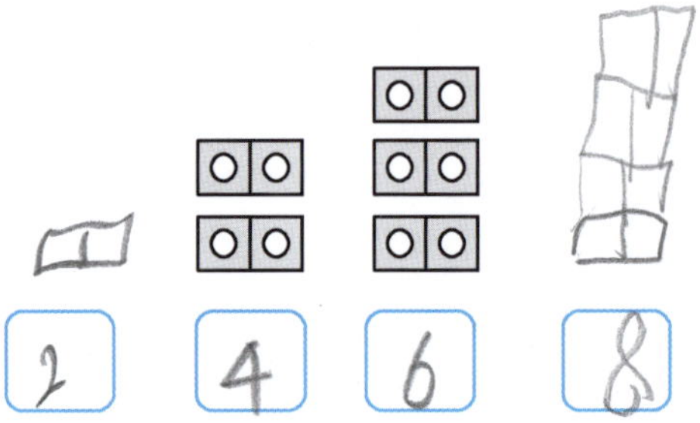

5. What numbers are next?

8, 10, 12, 14, 16

24, 26, 28, 30, 32

50, 48, 46, ~~44~~, 42

6. What number comes next on the number trains?

100, 98, 96, 94

4, 2, 0

1. Investigate the statements.
Do you agree?

If you count in steps of 2 from 0, the last digit of your number will be even.

Malachi

Yes

Explain or prove your answers here.

2. Which one is the odd one out?
Explain your reasons.

- 60, 62, 64, 67, 68, 70

70 because it has only 7 in it.

Explain or prove your answers here.

3.

I count backwards in steps of 2 from the number 16.
I count 6 times. The number I stop at is ________.

What number did Malachi stop at?

Explain or prove your answers here.

Number and Place Value

Count in steps of 3

1. Count in steps of 3 from the given number.

39					
0					
21					

2. Leanna has 6 football cards.

She collects 3 more every day.

Complete the number track to show how many she will have in 6 days.

6						

She will have _______ in 6 days.

3. Continue the pattern on the caterpillar.

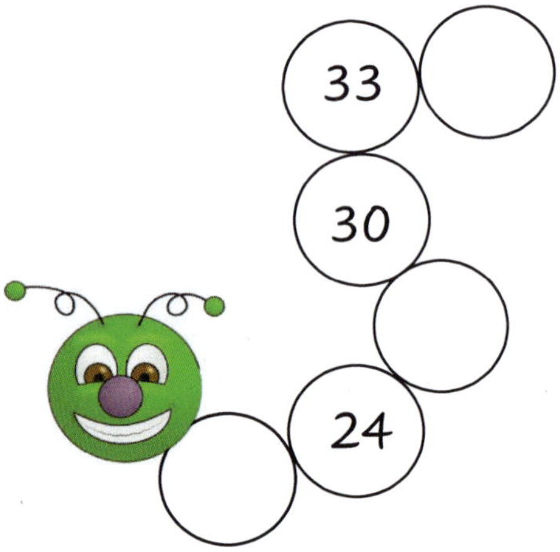

4. Complete the number sequence.

	9		3	

5. What numbers are next?

84, 87, 90, 93 , 96

33, 36, 39, 42 , 45

15, 18, 21, 24 , 27

6. Tia has 15 football cards.

She collects 3 more every day.
Complete the number track to show how many she will have in 4 days.

She will have _______ in 4 days.

Number and Place Value ★★★

Count in steps of 3

1. Investigate Rosie's statement.
Do you agree?

If you count in steps of 3 from 0, the last digit of your number will always be odd.

Rosie

Explain or prove your answers here.

2. Spot the mistake in the sequence and correct it.

0, 3, 6, 9, 11, 15

57, 60 , 61, 66, 69

Explain or prove your answers here.

3. Do you agree with Esin? Prove it!

Esin says:

If I start at 36 and count in 3s, I say the number 6.

Explain or prove your answers here.

1. Count in steps of 5 from the given number.

15 ☐ ☐ ☐ ☐ ☐

40 ☐ ☐ ☐ ☐ ☐

75 ☐ ☐ ☐ ☐ ☐

2. What comes before and after in the number sequence? Write the numbers in the box and draw the rest of the sequence.

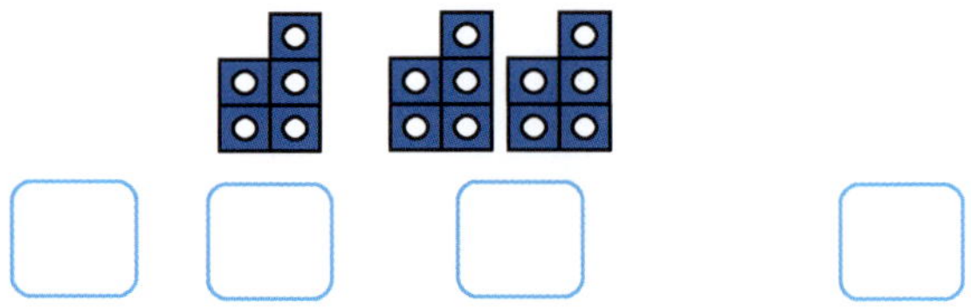

3. Continue the pattern on the caterpillar.

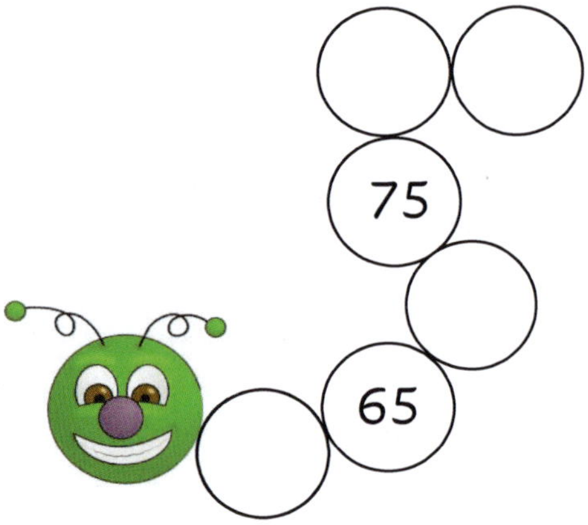

4. How much money is here?

☐ p

5. What numbers are next?

45, 50, 55, ____, ____

100, 95, 90, ____, ____

30, 25, 20, ____, ____

6. What number sequence is represented?

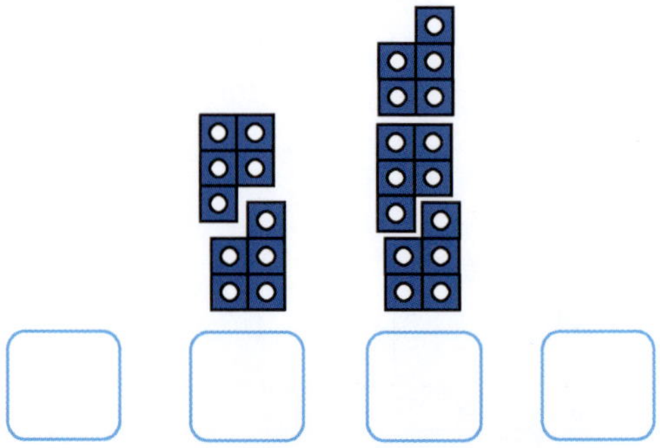

Number and Place Value

Count in steps of 5

1.

I count in 5s backwards from 20. I will say the number 10.

Zach

Explain or prove your answers here.

2. Always, Sometimes or Never?

When counting in 5s from zero the numbers will end in 0 or 5.

Explain or prove your answers here.

3.

I count in fives from zero up to 45.

I will say 9 numbers.

Malachi

Explain or prove your answers here.

Number and Place Value

Count in tens from any number, forward and backward

1. Write the number which is ten more.

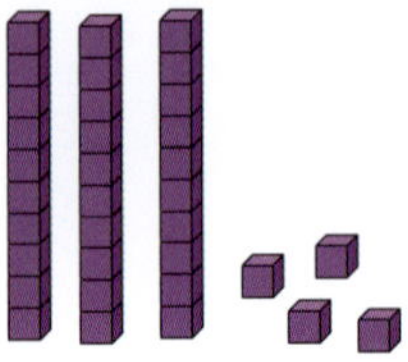

+ 10 →

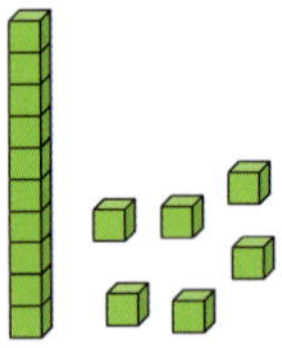

+ 10 →

2. Write the number which is ten less.

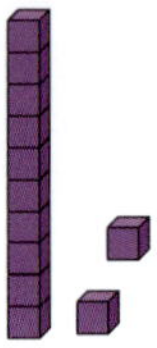

← – 10

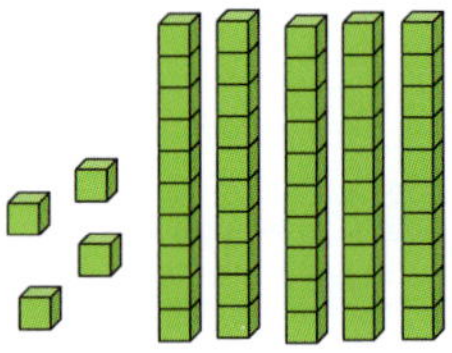

← – 10

3. What will the next three numbers be?

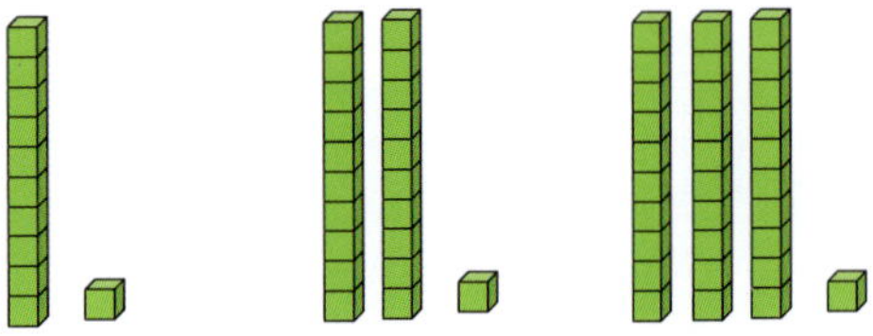

Next three numbers:

4. What will the next three numbers be?

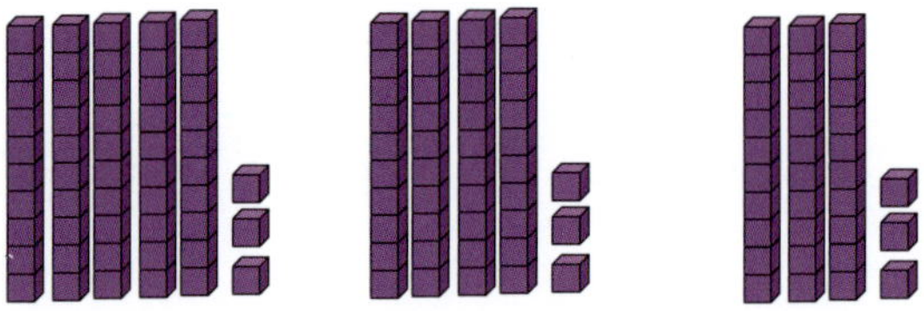

Next three numbers:

5. Complete.

8 tens and 6 ones = _______

7 tens and 6 ones = _______

6 tens and 6 ones = _______

____ tens and 6 ones = _______

____ tens and ____ ones = _______

____ tens and ____ ones = _______

6. Complete.

1 ten and 0 ones = _______

2 tens and 0 ones = _______

3 tens and 0 ones = _______

____ tens and 0 ones = _______

____ tens and ____ ones = _______

____ tens and ____ ones = _______

7. Write the number which is ten more.

12 → ☐

54 → ☐

89 → ☐

7 → ☐

8. Count in steps of 10 forward from the given number.

0			
25			
44			

9. Write the answer in the box.

17 + 10 = ☐

47 + 10 = ☐

43 – 10 = ☐

90 + 10 = ☐

10. Continue the pattern on the caterpillar.

11. What numbers comes next?

12. What comes before and after in the number sequence?

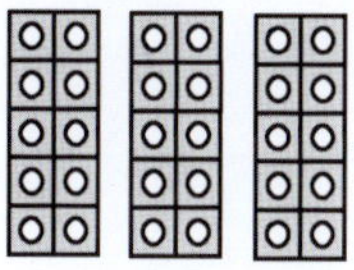

☐ ☐

☐ ☐

Number and Place Value

Count in tens from any number, forward and backward

1. Can you complete this using your own numbers?

☐ is 10 less than → ☐

☐ is 10 more than → ☐

Explain or prove your answers here.

2. Spot the mistakes.

100, 90, 80, 75, 60, 50, 45

43, 34, 44, 54, 64, 74, 84

28, 38, 48, 58, 68, 88, 89

Explain or prove your answers here.

3. Tia and Malachi are counting from 0 to 30.

Tia is counting in 10s.
Malachi is counting in 5s.

Malachi Tia

Will they say any of the same numbers?
What do you notice about your answer?

Explain or prove your answers here.

Number and Place Value
Recognise the place value of each digit in a two-digit number

1. How many tens and ones are there in the following numbers?

 63 ― tens and ― ones.
 90 ― tens and ― ones.
 21 ― tens and ― one.
 30 ― tens and ― ones.

2. Circle the numbers between 60 and 80.

 34
 4
 65
 74
 45
 23
 80
 61

3. Circle the numbers with 4 ones.

 34
 4
 74
 43 45
 43 14

4. What is the value of the number underlined?

 73 _____
 99 _____
 3**7** _____
 4**1** _____

5. What does the number 7 represent in these numbers? Ones or tens?

 67 _____ 7 _____
 70 _____ 97 _____

6. Break down these numbers into tens and ones.

 58 → ☐ ☐
 31 → ☐ ☐

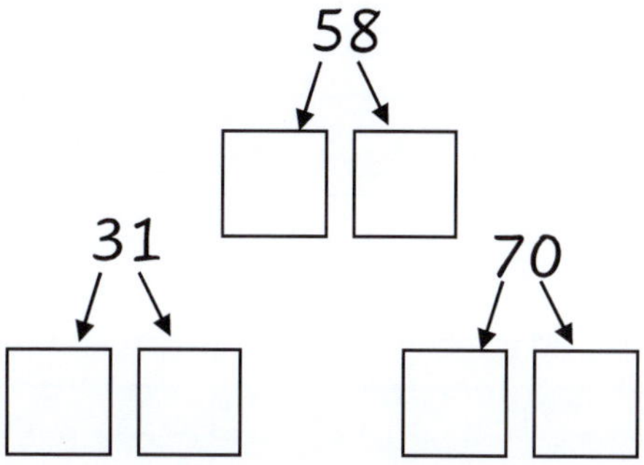

Number and Place Value

Recognise the place value of each digit in a two-digit number

7. Complete the part whole models.

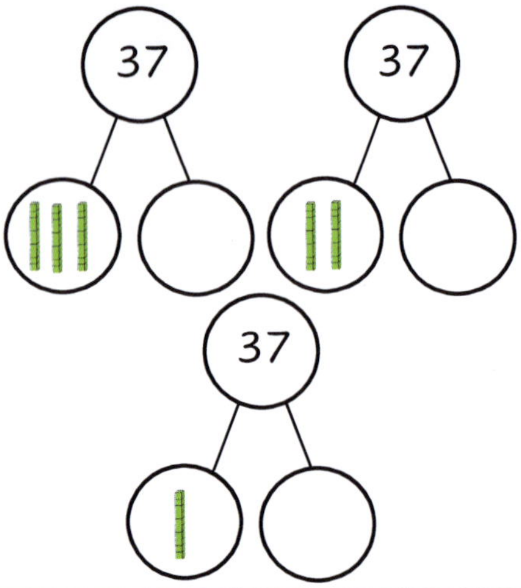

8. How many ways can you partition the following number?

9. Match the number sentences to the correct number.

34 24 44 14

30 + 4 20 + 14 10 + 24

10. Write a number sentence to show the number partitioned.

37 83 99

11. Fill in the missing numbers.

10 + 2 = 12

20 + ____ = 22

30 + 2 = ____

____ + 2 = 42

12. Fill in the missing numbers.

1 ten + 2 ones = 12

2 tens + ____ ones = 22

3 tens + 2 ones = ____

____ tens + 2 ones = 42

Number and Place Value ★★★

Recognise the place value of each digit in a two-digit number

1. True or False?
Explain your answer.

39 has more ones than 98.

Explain or prove your answers here.

2. Can you represent this number in a different way?

My number has 2 ones and 9 tens. What number am I thinking of?

Explain or prove your answers here.

3. Sort the numbers.

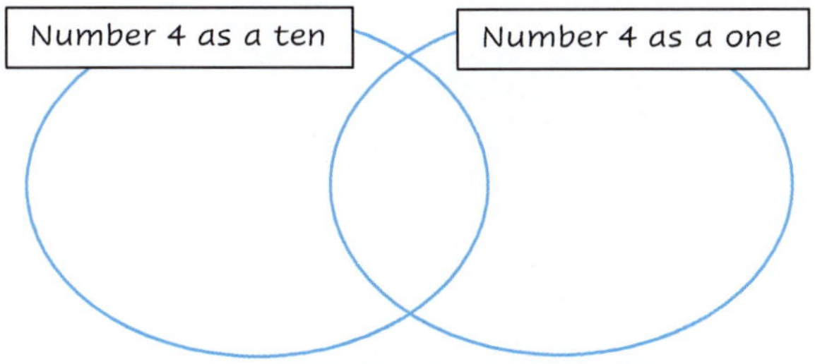

45 | 24 | 44 | 94 | 4 | 40 | 04

Explain or prove your answers here.

Number and Place Value

Identify numbers using different representations

1. Write the missing numbers on the number line.

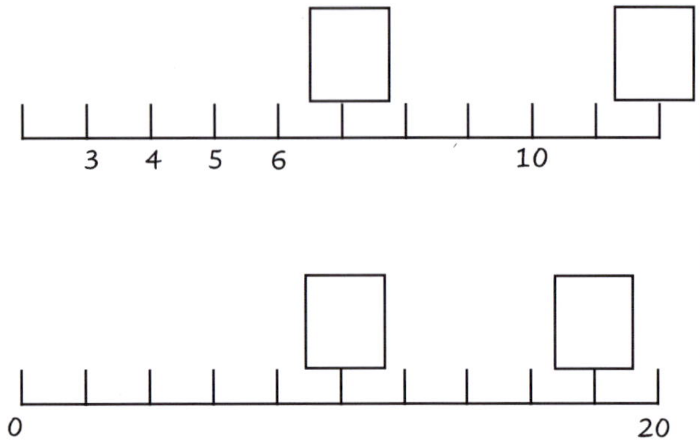

2. What numbers do the circles represent?

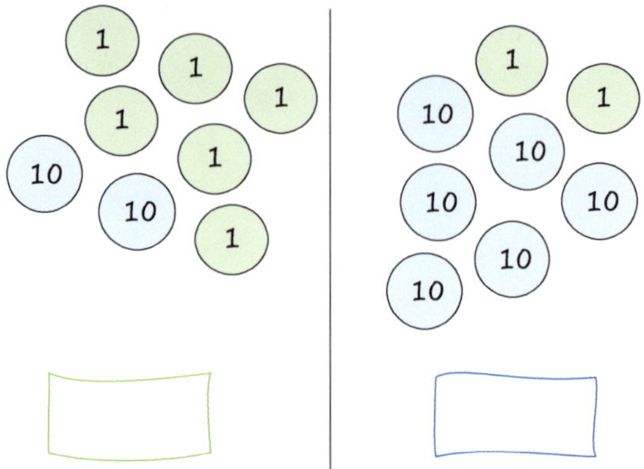

3. What number does the table represent?

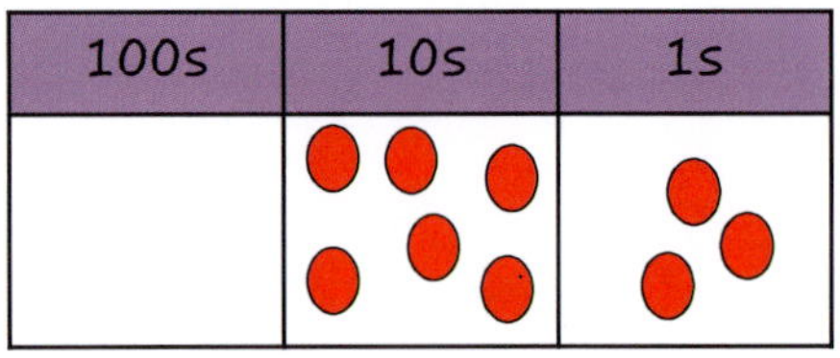

4. Use a drawing to represent the number 36 in its tens and ones.

5. What numbers do the cubes represent?

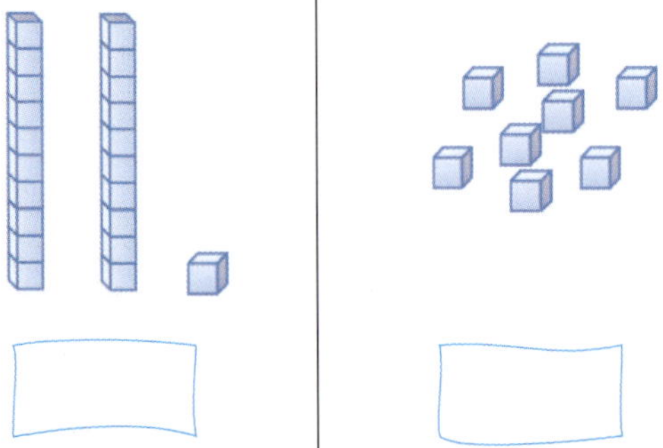

6. Place the number 10 on the number lines below.

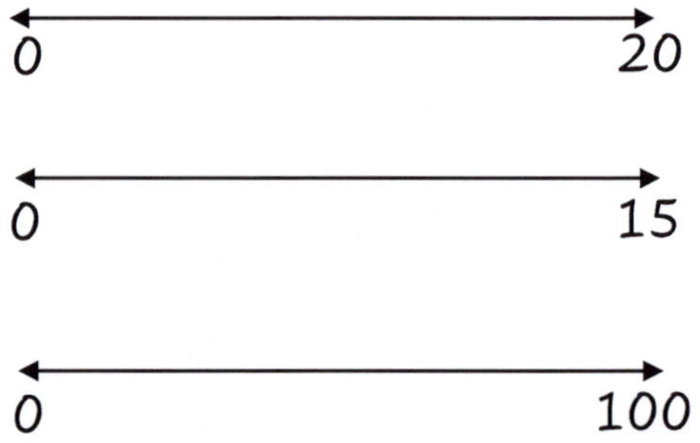

7. Here is part of a bead string.
Complete the sentences.

There are _______ tens
and _______ one.

The number is _________ .

8. What numbers are represented?

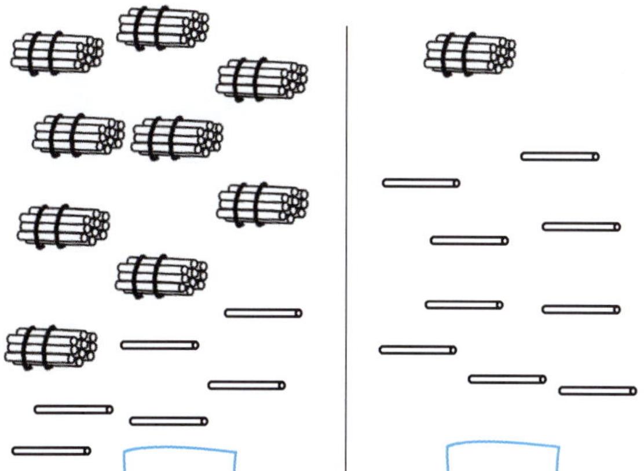

9. How many pencils?

10. Match the number to the correct representation.

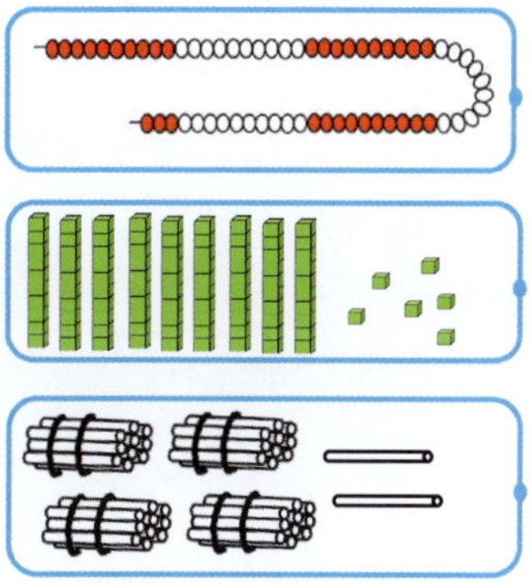

four tens and two ones

Sixty-three

nine tens and six ones

11. How many sweets?

12. What number is represented?
Write it in digits and words.

1. True or False?

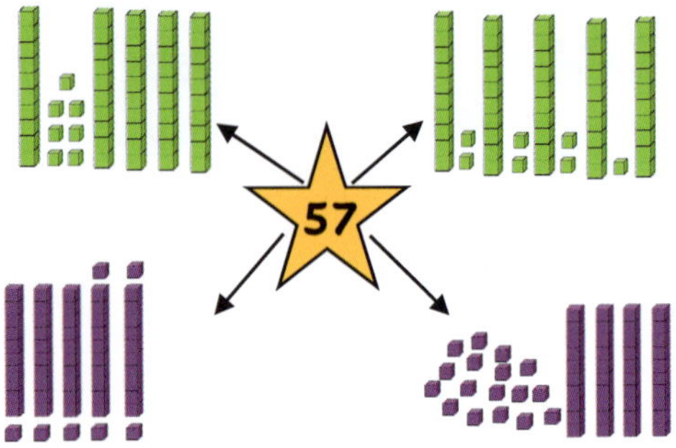

Explain your answer.

Explain or prove your answers here.

2. True or False?

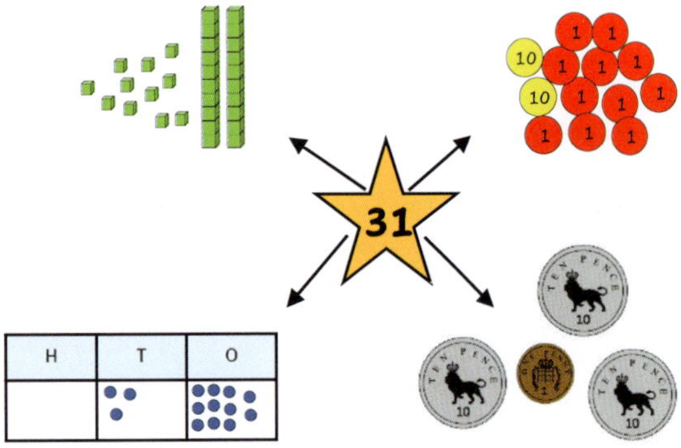

Explain your answer.

Explain or prove your answers here.

3. Tia and Esin think they are both correct. What do you think?

Tia says:

100 ones = 10 tens.

Esin says:

1 ten = 100 ones.

Explain or prove your answers here.

4. True or False?

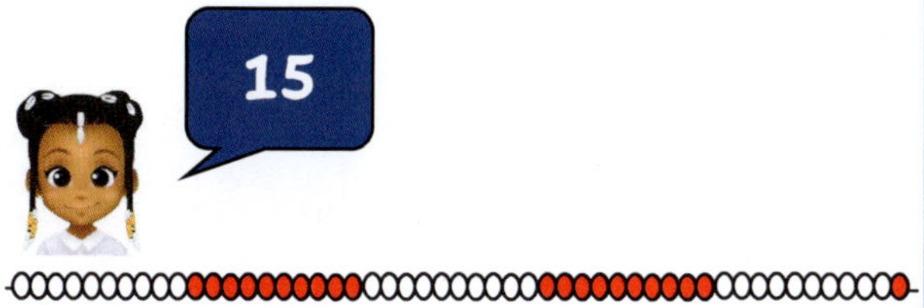

Tia says she has made the number 15.

Explain the mistake she has made.

Explain or prove your answers here.

5. Which is the odd one out? Explain.

A

There are two tens and one one.

B

There are twenty-one beads.

C

There are two ones and two tens.

Explain or prove your answers here.

6. One of these images **does not** show thirty-four. Which one is it?

A
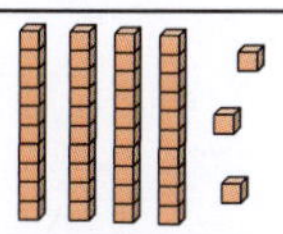

B
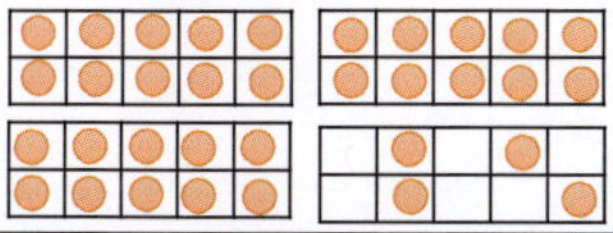

C
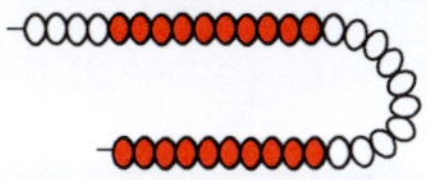

Explain or prove your answers here.

Number and Place Value

Compare numbers from 0 up to 100 using <, > and = signs

1. Add comparison symbols.

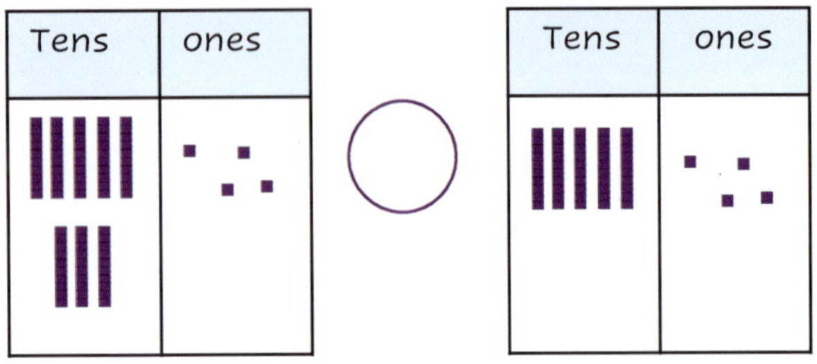

2. Add comparison symbols.

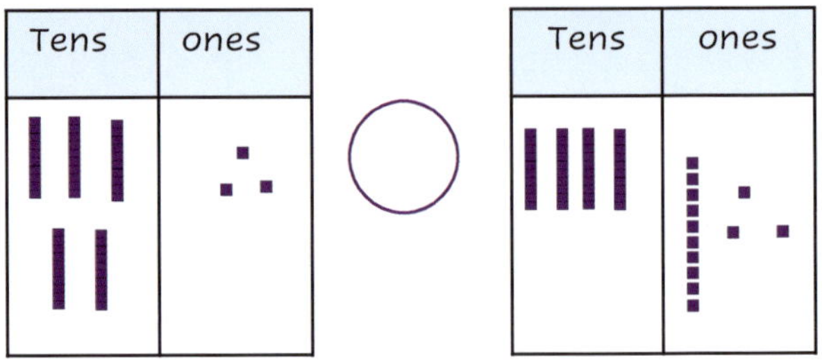

3. Use the signs below to make this true.

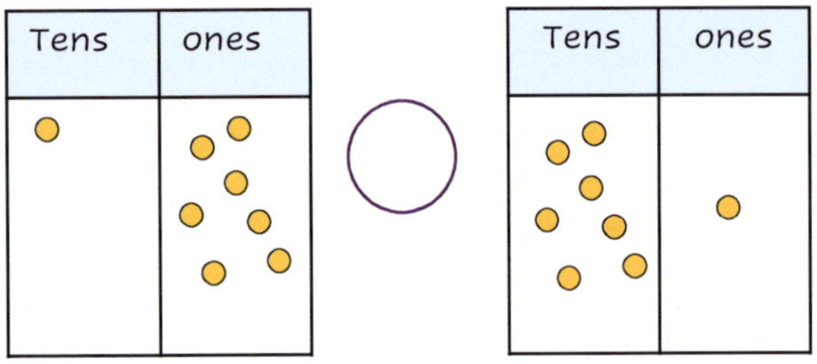

4. Compare the objects using comparison symbols.

> = <

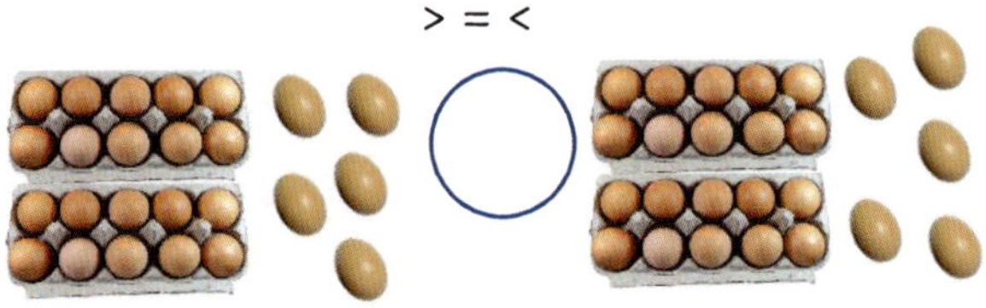

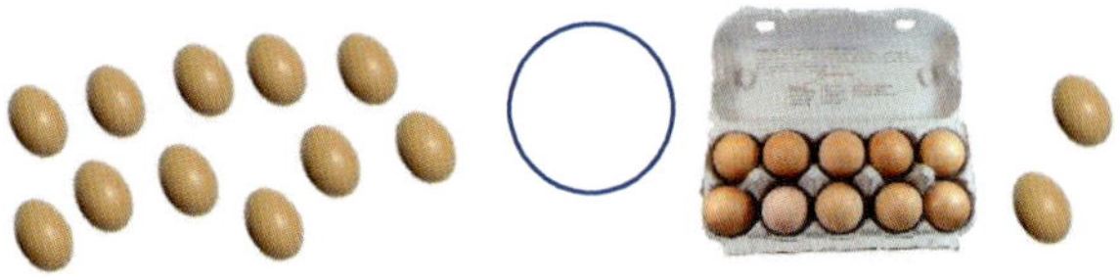

5. Add the words greater or less to make the statement correct.

63 is ____________________ than 89

19 is ____________________ than 91

62 is ____________________ than 59

81 is ____________________ than 18

6. Use the signs below to make these true.

Tens	ones
7	8

Tens	ones
8	7

Tens	ones
1	3

Tens	ones
1	9

Compare numbers from 0 up to 100 using <, > and = signs

7. Circle the smallest number in each row.

35	97	13	63
100	5	55	34
34	43	24	23

8. Circle the largest number in each row.

46	83	95	14
76	56	29	50
28	82	88	22

9. Use the signs below to make this true.

< = >

31 ☐ 64

98 ☐ 99

42 ☐ 42

10. Use the signs > < = to make this true.

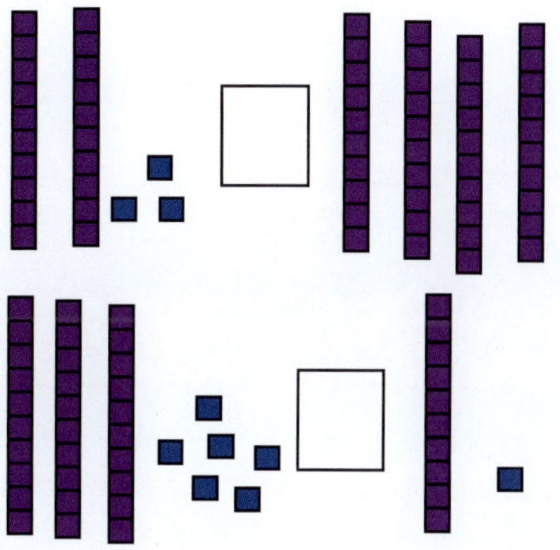

11. Use the signs below to make this true.

< = >

4 tens and 2 ones ☐ 42

2 tens and 4 ones ☐ 42

71 ☐ 6 tens and 1 one

12. Use the signs below to make this true.

< = >

3 tens and 2 ones ☐ 2 tens and 12 ones

7 tens and 1 ones ☐ 7 tens and 13 ones

6 tens and 4 ones ☐ 6 tens and 11 ones

Number and Place Value

Compare numbers from 0 up to 100 using <, > and = signs

1. Look at the numbers below:

40 | 72 | 0 | 19

Write each number once to make these correct.

☐ > ☐

☐ > ☐

Explain or prove your answers here.

2. Add a different number each time to make this correct.

10 < ☐

10 < ☐

10 < ☐

Explain or prove your answers here.

3. Zach and Esin are comparing numbers they have made.

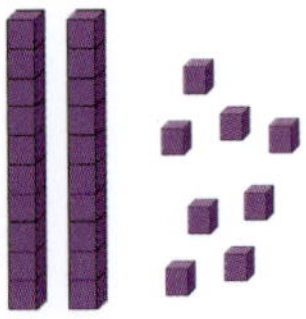

Esin's number

Zach's number

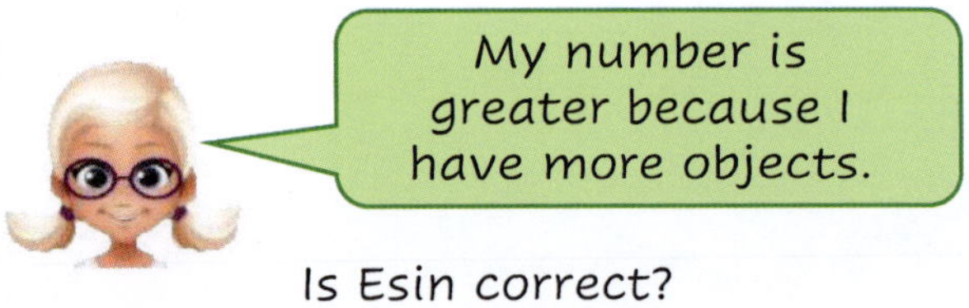

My number is greater because I have more objects.

Is Esin correct?
Explain your answer.

Explain or prove your answers here.

Number and Place Value

Order numbers from 0 up to 100

1. Write these numbers in order from the smallest to the largest.

43	34	44	33

2. Write these numbers in order from the largest to the smallest.

17	7	71	77

3. Label the numbers 1 – 5. 1 being the smallest.

Sixty-eight

Twenty-five

18

fifteen

50

4. Label the numbers 1 – 5. 1 being the **largest**.

fifty

fourteen

77

60 + 3

15

5. Label the teddies in order of price from the highest to the lowest.

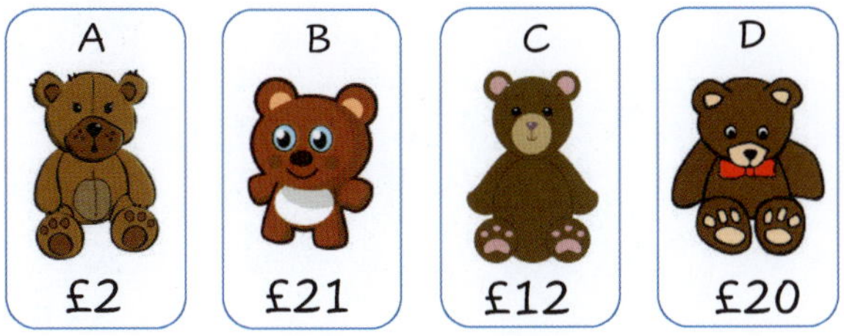

6. The children took part in a marathon! Order them from last to first.

Number and Place Value ★★★

Order numbers from 0 up to 100

1. Both children have ordered the numbers below. Who is correct?

81	7	44

Zach: 7, 44, 81

Esin: 81, 44, 7

Explain or prove your answers here.

2. Look at the numbers. Which one will be 4th if you are ordering them from the smallest to greatest ?

seven

71

7 + 1

17

seventy

Explain or prove your answers here.

3. These numbers have been ordered from greatest to smallest.

Twenty-one

eighteen

four

zero

True or false?
Explain your answer.

Explain or prove your answers here.

Number and Place Value

Read and write numbers to at least 100 in numerals and in words

1. Which numbers are spelt incorrectly? Circle them and correct them.

One hundred

ten

fifteen

eleven

fourty

ninty

eihgt

fiftee

2. Write the numbers in words.

35

76

12

9

44

3. Match the following numbers.

seventy-eight | seventy-four | sixty- seven

4. Write the numbers in digits.

thirty-two

ninety-eight

seventy-five

eighty

thirteen

5. Write the numbers in words.

Write the numbers between forty and sixty that end in zero and five.

6. Find the hidden numbers.

q	b	v	t	c	x	w	z	t	w	o
g	e	w	o	h	j	k	l	z	x	q
m	l	n	o	e	b	e	v	r	c	t
q	e	w	e	n	i	n	e	r	t	n
o	v	i	y	u	u	g	i	o	y	i
a	e	i	h	g	t	s	h	d	f	g
p	n	a	g	h	s	j	k	t	d	h
l	g	z	x	c	f	x	f	d	s	n
n	t	y	m	u	i	o	l	p	a	k
r	h	e	j	s	e	l	e	v	i	n

9

11

6

8

2

7. What number is represented below?

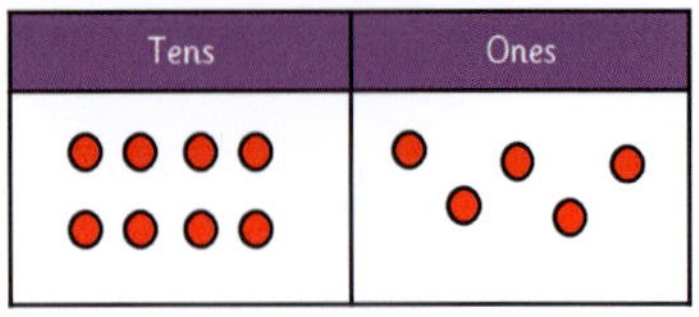

Numeral: __________

Words:______________________________

8. Match the numerals to the words.

37 | 83 | 99 | 16

ninety-nine | sixteen | thirty-seven | eighty-three

20 | 82 | 73 | 13

Seventy-three | eighty-two | twenty | thirteen

9. Tia is in a shop.

What can she buy for twenty-three pounds?

10. Write the numbers in digits.

seventeen ☐

thirty ☐

forty-one ☐

two ☐

one hundred ☐

11. Write the numbers in words.

89 ☐

58 ☐

42 ☐

33 ☐

7 ☐

12. Each jar contains ten marbles.

How many marbles are there altogether?

Write your answer in numerals and words.

☐

☐

Number and Place Value
Read and write numbers to at least 100 in numerals and in words

★★★

1. Three numbers have been written on a board.

 | forty | forteen | for |

 Which one is the odd one out?
 Explain your answer.

 Explain or prove your answers here.

2. What is the same and what is different about the numbers below?

 40 14 41 4

 Write the numbers in words when you explain your answer.

 Explain or prove your answers here.

3. Tia says:

 Thirty-seven add thirty-two equals sixty-nine

 She also writes it on her notepad.
 Are they the same?
 Is she correct?

 $37 + 32 = 69$

 Explain or prove your answers here.

Number and Place Value

Use place value and number facts to solve problems

1. Which alien holds the box that will balance the scales?

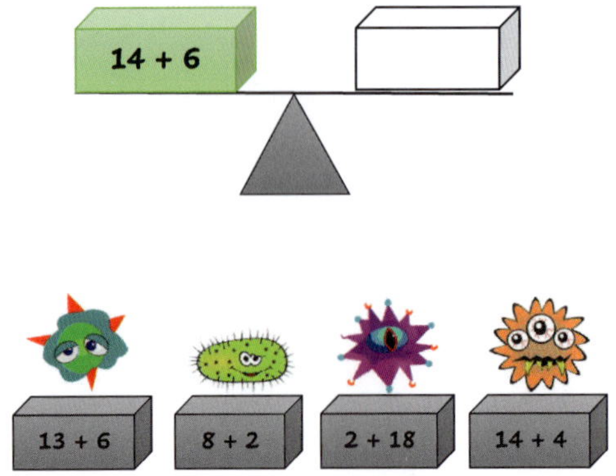

2. I caught 20 pokemon.
I wanted to catch 30.
How many more do I need to catch?

3. I have 4 marbles in one jar.
How many marbles do I have to put in the other jar to make 12 altogether?
Draw them in.

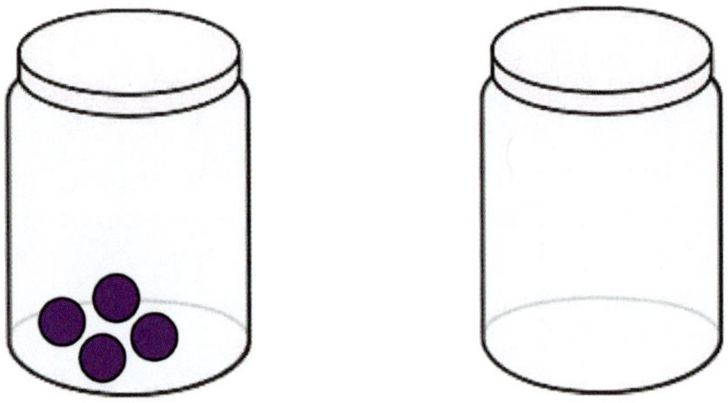

4. The numbers in the triangle add up to fourteen.
Write the missing number.

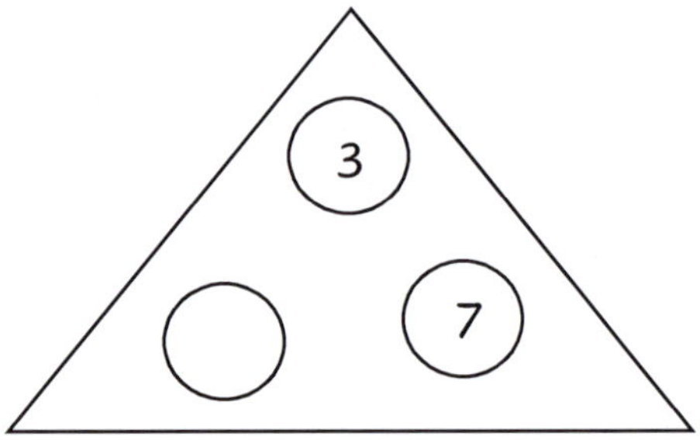

5. Farmer Joan needs to have 50 eggs altogether. Write the number of eggs more that she needs on the basket.

6. Is this correct? If not, what can you do so that it is correct?

Yes ☐ No ☐

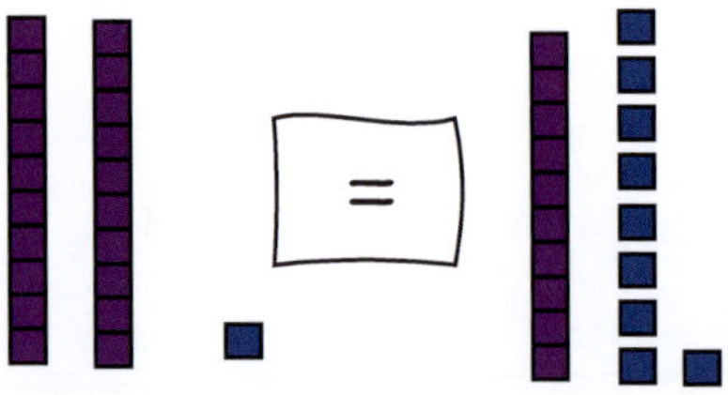

Number and Place Value ★★★

Use place value and number facts to solve problems

1. According to the pattern, write in words what the next price should be.

These items are for sale:

£12 | £ 22 | £32

Explain or prove your answers here.

2. Does Rosie have enough money?

I have 9 ten-pound notes and 8 one-pound coins. I would like to buy this game.

Explain or prove your answers here.

3. What could her number be?

Leanna says:

I am counting backwards and start at the number 48. I say five odd numbers and five even numbers.

Could it be more than one number?

Explain or prove your answers here.

Number and Place Value

Colour in the amount of stars you got correct for each section.

Objective	Fluency	Reasoning
Count in steps of 2	☆☆☆☆☆☆	☆☆☆
Count in steps of 3	☆☆☆☆☆☆	☆☆☆
Count in steps of 5	☆☆☆☆☆☆	☆☆☆
Count in tens from any number, forward and backward	☆☆☆☆☆☆ ☆☆☆☆☆☆	☆☆☆
Recognise the place value of each digit in a two-digit number	☆☆☆☆☆☆ ☆☆☆☆☆☆	☆☆☆
Identify numbers using different representations	☆☆☆☆☆☆ ☆☆☆☆☆☆	☆☆☆ ☆☆☆
Compare numbers from 0 up to 100 using <,> and = sign	☆☆☆☆☆☆ ☆☆☆☆☆☆	☆☆☆
Order numbers from 0 up to 100	☆☆☆☆☆☆	☆☆☆
Read and write numbers to at least 100 in numerals and in words	☆☆☆☆☆☆ ☆☆☆☆☆☆	☆☆☆
Use place value and number facts to solve problems	☆☆☆☆☆☆	☆☆☆

Do you know what these words mean? Tick the words that you know.
Go back to the beginning to revise the words you do not know.

Place Value		Less/Fewer/Fewest		Part-Whole Model		Sequence	
Digit		More/Greater/Greatest		Equal		Pattern	
Tens and Ones		Compare		Before		Even Numbers	
Ten Frames		Representation		After		Odd Numbers	
Numeral		Inequality Symbols		Next		Hundreds	
Number Track		Partition		Between		Columns and Rows	

Answers

Count in steps of 2

1. 6,8,10,12,14. 22,24,26,28,30. 50,52,54,56,58
2. 18, 12
3. 28, 30, 32, 34, 36, 38, 40, 42
4. 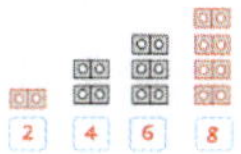

5. 14,16. 30,32. 44,42.
6. 94. 0

Reasoning

1. I agree. All the multiples of 2 end in an even number. Children can list out the numbers and prove this.

2. 67 because this number does not appear if we are counting in 2s from 0.

3. The number 4. Children can write out this sequence to find the number.

Count in steps of 3

1. 42,45,48,51,54. 3,6,9,12,15 24,27,30,33,36
2. 9, 12, 15, 18, 21, 24
3. 21 ,24, 27, 30, 33, 36
4. 12, 9, 6, 3, 0
5. 93, 96. 42, 45 24, 27
6. 18, 21, 24, 27. She will have 27 football cards.

Reasoning

1. I do not agree. Children can write out a sequence of 3s. 0,3,6,9,12,15. In the sequence, you can see that 6 is an even number and 12 ends in 2. This is an even number.

2. 11 should be 12. 61 should be 63.

3. This depends if Esin is counting forwards or backwards. If she is counting forwards, then I would not agree. If she is counting backwards, then I agree because the number sequence would look like this:

36, 33, 30, 27, 24, 21, 18, 15, 12, 9, 6

Count in steps of 5

1. 20, 25, 30, 35, 40. 45,50, 55, 60, 65. 80, 85, 90, 95, 100
2. 

3. 60, 65, 70, 75, 80, 85
4. 65
5. 60, 65. 85, 80. 15, 10
6. 5, 10, 15, 20

Reasoning

1. True. The sequence can be shown to prove this. 20, 15, 10

2. Always. 0, 5, 10, 15, 20, 25, 30 ...

3. False, he will say 10 numbers if he includes 0. 0, 5, 10, 15, 20, 25, 30, 35, 40, 45

Count in tens from any number, forward and backward

1. 34 + 10 = 44 16 + 10 = 26
2. 12 – 10 = 2 54 – 10 = 44
3. 11, 22, 33, 44, 55, 66
4. 53, 43, 33, 22, 11, 1
5. 86, 76, 66, 5 tens and 6 ones = 56, 4 tens and 6 ones = 46, 3 tens and 6 ones = 36
6. 10, 20, 30. 4 tens and 0 ones = 40, 5 tens and 0 ones = 50, 6 tens and 0 ones = 60
7. 22, 64, 99, 17
8. 10, 20, 30. 35, 45, 55. 54, 64, 74
9. 27, 57, 33, 100
10. 97, 87, 77, 67, 57, 47, 37, 27
11. 32, 42
12. 20, 30, 40, 50

Reasoning

1. Various answers. 4 is ten less than 14/ 56 is 10 more than 46.

2.
   - 100, 90, 80, 75, 60, 50, 45
   - 43, 34, 44, 54, 64, 74, 84
   - 28, 38, 48, 58, 68, 88, 89

3. They will both say 0, 10, 20 and 30.

Answers

Recognise the place value of each digit in a two-digit number

1. 6 tens and 3 ones, 9 tens and 0 ones, 2 tens and 1 one, 3 tens and 0 ones.
2. 65, 74, 61
3. 34, 4, 74, 14
4. 7 tens, 9 tens, 7 ones, 1 one
5. 67- ones 7- ones 70- tens 97- ones
6. 50,8 30, 1 70, 0
7. 7, 17,27
8. Various answers: 70 + 2, 60 + 12, 50 + 22, 40 + 32...
9. 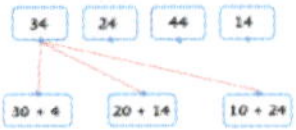
10. 37 = 30 + 7. 83 = 80 + 3. 99 = 90 + 9
11. 20 + 2 = 22, 30 + 2 = 32, 40 + 2 = 42
12. 2 tens + 2 ones = 22, 3 tens + 2 ones = 32
4 tens + 2 ones = 42

Reasoning

1. Children may say true as there are 9 ones in the 9 place and 8 ones in the ones place for 98. Some children will recognise that 98 could be made from 98 ones so therefore have more ones than 39. This is open for discussion.

2. The number is 92. This can be represented in many ways. 92 ones. 82 ones and 1 ten etc.

3. 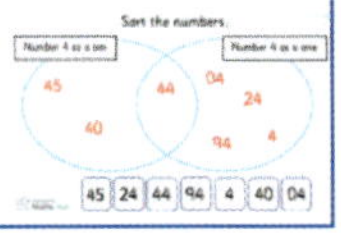

Identify numbers using different representations

Fluency

1. 7, 12. 10,18
2. 26, 62
3. 63
4. Various drawings: 
5. 21, 8
6. 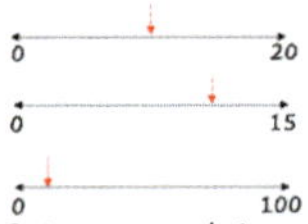
7. 3 tens and 1 one. 31
8. 96, 19
9. 62
10. 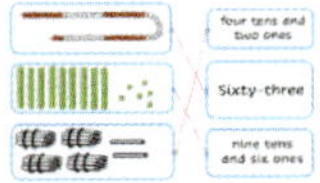
11. 65
12. 32, thirty-two

Reasoning

1. True. All the base 10 show 57.
2. False, the place value chart shows 41, not 31.
3. Tia is correct and Esin is not. 1 ten = 10 ones.
4. Tia has made the number 51 instead. She has mistaken her tens for ones.
5. Possible answer: C because it shows 22 and the others show 21.
6. A shows 43, not 34.

Compare numbers from 0 up to 100 using <,> and = signs

1. >. (84 > 54)
2. = (53 = 53)
3. < (17 < 71)
4. 25 = 25. 11 < 12
5. less. less. more. more
6. 78 < 87. 13 < 19
7. 13, 5, 23
8. 95, 76, 88
9. 31 < 64. 98 < 99 42 = 42
10. 23 < 40. 36 > 11
11. = , < , >
12. = , <, <

Reasoning

1. Example answers: 72 > 40. 19 > 0 40 > 0. 72 > 19
2. Various answers. Any number more than 10. 11,12,13
3. False. Esin has more one pieces; this does not mean she has a greater number. Her number is 28 and Zach's number is 41 so Zach's is greater.

Order numbers from 0 up to 100

1. 33, 34, 43, 44
2. 77, 71, 17, 7
3. 
4. 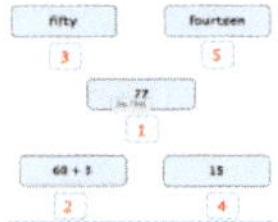
5. B,D,C,A.
6. B,D,C,A

Reasoning

1. This depends on how they are ordering the numbers. If they are ordering from smallest to largest then Zach is correct. If they are ordering from greatest to smallest, then Esin is correct.
2. Seventy.
3. True.

Answers

Read and Write numbers to at least 100 in numerals and in words

1. Ninty- ninety. Fourty- forty. Eihgt- eight
   fiftee- fifty
2. Thirty-five, seventy-six, twelve, nine, forty-four
3. 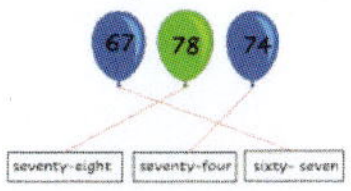

4. 32, 98, 75, 80, 13
5. Forty-five, fifty, fifty-five
6. 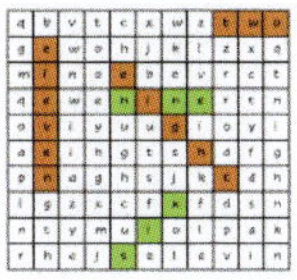
7. 85, eighty-five
8. 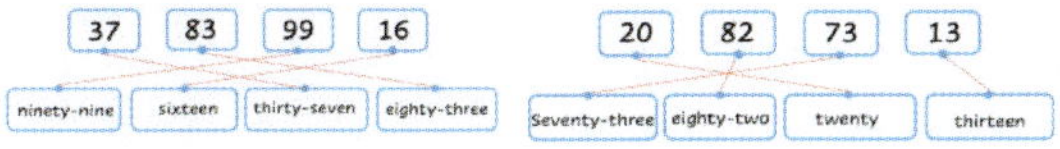

9. The ball because it is twenty-three pounds.
10. 17, 30, 41, 2, 100
11. Eighty-nine, fifty-eight, forty-two, thirty-three, seven
12. 46, forty-six

Reasoning

1. The last board because it has been written incorrectly. It says for, when it should say four.
2. Possible answer: They all have the number four but they are worth different amounts. 4 has one digit, the others have two digits.
3. Tia is correct, they are both the same.

Use place value and number facts to solve problems

1. 20- third alien
2. 10 more.
3. 8 circles drawn in.
4. 4
5. 40
6. No, 2 more are needed.

Reasoning

1. Forty-two pounds.
2. Yes, she does have enough money. 9 tens and 8 ones= 98. She has £98 and the game is £89
3. Her number could be 38 because:
   48 --- 47, 46, 45, 44, 43, 42,
   41, 40, 39, 38

   She could also be counting in 3s backwards.
   48, 45, 42, 39, 36, 33,
   30,27,24, 21, 18

Remember to check out our website for lots of free resources:

www.masterthecurriculum.co.uk

Printed in Great Britain
by Amazon

28893639R00025